WITHDRAWN

The Rain Forest

Plants of the Rain Forest

Mae Woods
ABDO Publishing Company

visit us at
www.abdopub.com

Published by Abdo Publishing Company 4940 Viking Drive, Edina, Minnesota 55435.
Copyright © 1999 by Abdo Consulting Group, Inc. International copyrights reserved in all countries. No part of this book may be reproduced in any form without written permission from the publisher.

Printed in the United States.

Photo credits: Peter Arnold, Inc.

Edited by Lori Kinstad Pupeza
Contributing editor Morgan Hughes
Graphics by Linda O'Leary

Library of Congress Cataloging-in-Publication Data

Woods, Mae.
   Plants of the rain forest / Mae Woods.
   p. cm. -- (Rain forest)
   Includes index.
   Summary: Introduces the plants of the rain forest, discussing both the forest floor and the treetops and examining how plants interact with the animals and provide food and other products.
   ISBN 1-57765-018-2
   1. Rain forest plants--Juvenile literature. 2. Rain forest ecology--Juvenile literature. [1. Rain forest plants. 2. Rain forest ecology. 3. Ecology.] I. Title. II. Series: Woods, Mae. Rain forest.
QK938.R34W66   1999
581.7'34--dc21   97-53101
                                                                             CIP
                                                                             AC

<u>Note to reader</u>
The words in the text that are the color green refer to the words in the glossary.

Contents

The Rain Forest 4
Layers of Life 6
The Forest Floor 8
The Understory 10
The Canopy 12
Tree Tops and Tree Roots 14
How Animals Help Plants Grow 16
Food from the Rain Forest 18
Products Made from Plants 20
Glossary 22
Internet Sites 23
Index 24

The Rain Forest

The rain forest is warm and moist all year. This makes it an ideal place for plants to grow. More species of plants live in tropical rain forests than any other place in the world.

There is little seasonal change in rain forests, so plants are always green and do not lose their leaves. Plants and trees grow very quickly in rain forests.

The floor of a rain forest is home to millions of ants, spiders, termites, caterpillars, worms, and insects. Water animals such as frogs, turtles, and crocodiles live in its rivers. Birds, snakes, and monkeys live in its trees. Bats, deer, gorillas, and fierce jaguars and leopards live here, too.

All these different animals need the plant life in rain forests for food and shelter. The thick, leafy branches on the trees protect living things from the rain. It rains nearly every day.

Red-eyed tree frogs live among plants in rain forests.

Layers of Life

The rain forest always has shade at ground level and lots of sun above its tall trees. To study the many different types of plants living there, scientists divide the rain forest into four levels, like a building. Each separate layer becomes a unique habitat, or home, for different plants and animals.

The first level is called the forest floor. Rain forest plants allow very little sun to reach the ground. The next level, the understory, is a thick jungle. Trees living here are 15 to 60 feet (5 to 18 m) tall. Above this layer is the canopy. It is like an umbrella that blocks the rays of the sun. Plants in the canopy enjoy light all year round, and the trees reach

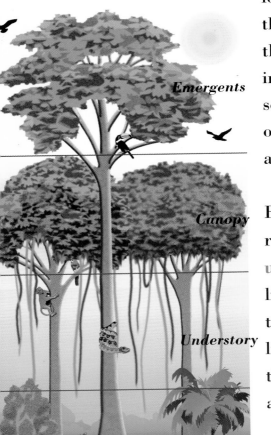

This illustration shows the different layers in a rain forest.

heights over 100 feet (30 m). Poking through the canopy are the tallest trees in the forest. Some are over 200 feet (61 m) tall. These are called emergents.

Plants in an Australian rain forest.

The Forest Floor

Ferns, mosses, and wild mushrooms grow on the floors of rain forests. These plants like the shade. Leafy plants and flowers cannot live here because they need sunshine in order to grow.

The forest floor is covered with leaves and branches that fall from the trees. The leaves quickly decay in the moisture. Then they are absorbed back into the soil as nutrients. They become food for plants and trees. Because of the heat and rain, tropical soils are not very rich or deep. All the nutrients are found in a thin layer of topsoil.

The dead trees and branches on the ground are eaten by termites. The termites are eaten by orangutans. When one plant or animal becomes food for another, it is part of a food chain. This chain begins with plant life and ends with the meat eating animals who eat other creatures.

Mushrooms and decomposing leaves litter the ground in the Amazon rain forest.

The Understory

The understory is a jungle of bushes and trees growing closely together. The canopy above blocks out part of the sun so there is not enough light for most flowers to bloom. However, many vines, ferns, palms, and bamboo thrive in the understory. Bamboo is actually a type of grass, but in rain forests it grows as thick and tall as the trees.

Woody vines called lianas are unique to this area. Lianas wrap themselves around tree trunks and climb upward to reach the sun. They grow from tree to tree, creating a rope of vines through the jungle. Monkeys can travel for miles very quickly by swinging from these strong vines.

Many plants in the understory have thick, waxy leaves with pointed ends called drip tips. When rain pours onto the leaf, the drip tip acts like a spout to drain away the water before it can damage the plant.

Trees and plants grow close together in the understory layer of a rain forest.

The Canopy

Most plants live in the sunny canopy level of a rain forest. This layer stretches from 50 to 160 feet (15 to 49 m) into the sky. It is thick with trees and vines. Tree tops grow close together and form an umbrella that keeps a lot of rain and sunlight from reaching the ground.

Here there are many flowering plants and mosses that can actually grow on trees and vines instead of in the soil. These "air plants" are called epiphytes. Their roots nest in the thin layer of compost formed on the branches. They live off nutrients found in air and water. Epiphytes, like orchids, often have beautiful flowers with sweet nectar for insects and birds to drink.

More animals live in the canopy than in any other part of a rain forest. It provides shelter and plenty of food. They can live their entire lives without ever seeing the ground!

The Monteverde Cloud Forest, Costa Rica.

Tree Tops and Tree Roots

The emergent level, made up of trees that grow to 200 feet (61 m), is the roof of a rain forest. These tree tops are always in the sun. Sometimes it is 10°F (6°C) hotter on this level than it is on the ground. On the ground the air is much more humid, which means it is more damp and moist.

Most emergent trees have bare trunks with thick, leafy branches at their crowns. They bloom in shades of yellow, red, and purple. Although they are huge, these trees have shallow roots. Rain forest soil is not very deep, so the trees grow buttress roots for extra support. These are a web of long roots at the base of the trunk. Trees would topple over without them. Vines help protect trees, too. They drape from one tree to another like a net and give support against the wind.

Opposite page: Tall rain forest trees have shallow roots.

How Animals Help Plants Grow

Plants and animals help each other grow. The fruits, vegetables, nuts, stalks, and leaves in rain forests can be food for many different animals. Trees and plants also shelter them and are good places for nests.

In turn, animals help plants grow by spreading seeds. Monkeys and apes eat fruit. They stuff the fruit in their mouths as they move through the rain forests. Fruit falls to the ground as the monkeys and apes move. New fruit trees will grow from the seeds of the fruit.

Many plants have spores or pollen instead of seeds. In some areas of the world, these are spread by the wind. In rain forests, where there is very little wind, insects and birds must do the job. Bees climb inside brightly colored flowers to drink their nectar and become coated in **pollen**. Then the bees fly to other flowers, bringing along the tiny grains that will grow new plants.

This orangutan helps spread seeds throughout the Malaysian rain forest.

Food from the Rain Forest

Tropical plants provide the world with many foods and spices. Ginger, nutmeg, peppercorns, and vanilla beans come from rain forests. Tapioca, used in dessert puddings, is a grain from the cassava root. Brazil nuts only grow in the damp forests of South America.

Many fruits were first found growing in rain forests, but are now farmed in other areas of the world. These include bananas, papaya, pineapples, oranges, peppers, and mangoes. Sugar, tea, coffee, and chocolate are made from plants that were first found here.

Chocolate comes from the seeds of the cacao tree. This tree bears colorful flowers that bloom and then turn into pods. Each pod contains rows of white seeds called cacao beans. The beans are removed, dried, and roasted. Then they are ground into powder and melted into a gooey syrup. This syrup is used to make chocolate candy and other delicious treats.

Banana trees grow in rain forests.

Products Made from Plants

Many products come from rain forest plants. The lianas vines that grow in the understory are as strong as wood. Their fiber, known as rattan, is used to make baskets and furniture. Jute is another fiber product made from plants. It is woven into burlap and also used to make rope, bags, and hammocks.

Giant trees from rain forests provide hardwood to make furniture. Teak, mahogany, balsa, and ebony are kinds of wood that come from trees in rain forests.

The kapok tree grows cotton-like seeds used for stuffing toys. The rubber tree contains a milky sap called latex, which is used to make rubber. Rubber farmers drill spouts into trees to drain the liquid. This does not hurt the trees, and they continue to make more latex. Medicine also comes from plants in rain forests. Scientists have found plants that can be turned into medicine and fight diseases. A plant called taxol might even be a cure for cancer.

Milky sap from rubber trees is used to make rubber.

Glossary

Canopy - the upper layer of a rain forest that has the most growth.
Compost - a mixture of decaying leaves.
Crown - the top of a tree.
Damage - hurt.
Decay - rot.
Drip tips - pointed ends of certain plant leaves.
Emergent - a tree that rises above the surrounding forest.
Epiphyte - a plant that grows on another live plant without damaging it.
Habitat - the place where an animal can survive.
Jute - rope-like fiber used to make burlap bags and hammocks.
Kapok - fibers that look like cotton used to fill mattresses and stuffed animals.
Latex - liquid that is used to make rubber.
Lianas - a woody vine.
Nectar - sweet liquid inside a flower.
Nutrients - food needed for life and growth of plants, animals, and people.
Pollen - a fine grain found inside a flower.
Rattan - fiber from lianas used to make baskets and furniture.
Sap - liquid that flows through a plant or tree.
Shelter - protection.
Species - a group of plants or animals that are alike in certain ways.
Spores - tiny cells that can grow into a new plant.
Topsoil - top level of the ground where the soil is rich in nutrients.
Understory - the middle level of the rain forest where trees and bushes grow.
Unique - special; one of a kind.

Internet Sites

Amazon Interactive
http://www.eduweb.com/amazon.html
Explore the geography of the Ecuadorian rain forest through on-line games and activities. Discover the ways that the Quichua live off the land.

Living Edens: Manu, Peru's Hidden Rain Forest
http://www.pbs.org/edens/manu/
This site is about the animals and indigenous people who populate Peru's Manu region.

The Rain Forest Workshop
http://kids.osd.wednet.edu/Marshall/rainforest_home_page.html
The Rain Forest Workshop was developed by Virginia Reid and the students at Thurgood Marshall Middle School, in Olympia, Washington. This site is one of the best school sites around with links to many other sites as well as great information on the rain forest.

The Tropical Rain Forest in Suriname
http://www.euronet.nl/users/mbleeker/suriname/suri-eng.html
A multimedia tour through the rain forest in Suriname (SA). Read about plants, animals, Indians, and Maroons. This site is very organized and full of information.

These sites are subject to change. Go to your favorite search engine and type in Rain Forest for more sites.

Pass It On

Rain Forest Enthusiasts: educate readers around the country by passing on information you've learned about rain forests. Share your little-known facts and interesting stories. Tell others about animals, insects, or people of the rain forest. We want to hear from you!

To get posted on the ABDO Publishing Company website E-mail us at
"Science@abdopub.com"
Visit the ABDO Publishing Company website at www.abdopub.com

Index

A

air plants 12
animals 4, 6, 8, 12, 16
ants 4

B

bamboo 10
bananas 18
beans 18
bees 16, 17
birds 4, 12, 16

C

cacao bean 18
canopy 6, 7, 10, 12
chocolate 18
compost 12
crown 14

D

diseases 21
drip tips 10

F

ferns 8, 10
flowers 8, 10, 12, 17, 18
food 4, 8, 12, 16, 18
forest floor 6, 8
fruits 16, 18

H

habitat 6

I

insects 4, 12, 16

L

latex 20, 21
leaves 4, 8, 10, 16

M

medicine 21
monkeys 4, 10, 16
moss 8, 12

N

nectar 12, 16
nutrients 8, 12
nuts 16, 18

P

products 20

R

roots 12, 14

S

sap 20
scientists 6, 21
seasonal change 4
shelter 4, 12, 16
snakes 4
soil 8, 12, 14
South America 18
species 4
spices 18
spiders 4
sun 6, 8, 10, 12, 14

T

trees 4, 6, 8, 10, 12, 14, 16, 18, 20
tropical 4, 8, 18

U

understory 6, 10, 11, 20

V

vegetables 16
vines 10, 12, 14, 20

W

water animals 4